Vaibhav Hudda, Akshay Kumar

# A Summary of Electric Vehicle Propulsion Technologies

GRIN Verlag

**Bibliografische Information der Deutschen Nationalbibliothek:**

Die Deutsche Bibliothek verzeichnet diese Publikation in der Deutschen National-
bibliografie; detaillierte bibliografische Daten sind im Internet über http://dnb.d-
nb.de/ abrufbar.

**Imprint:**

Copyright © 2014 GRIN Verlag GmbH
Druck und Bindung: Books on Demand GmbH, Norderstedt Germany
ISBN: 978-3-656-71018-9

**This book at GRIN:**

http://www.grin.com/en/e-book/277317/a-summary-of-electric-vehicle-propulsion-
technologies

**GRIN - Your knowledge has value**

Der GRIN Verlag publiziert seit 1998 wissenschaftliche Arbeiten von Studenten, Hochschullehrern und anderen Akademikern als eBook und gedrucktes Buch. Die Verlagswebsite www.grin.com ist die ideale Plattform zur Veröffentlichung von Hausarbeiten, Abschlussarbeiten, wissenschaftlichen Aufsätzen, Dissertationen und Fachbüchern.

**Visit us on the internet:**

http://www.grin.com/

http://www.facebook.com/grincom

http://www.twitter.com/grin_com

# A Summary of Electric Vehicle Propulsion Technologies

**Vaibhav Hudda, Akshay Kumar**

Department of Electrical and Electronics

AKGEC Ghaziabad, INDIA

*Abstract*— Electric vehicles (EV) firstly introduced in late 1800's. But the benefits offered by internal combustion engines over electric propulsion made the previous a popular choice. Increasing price of fossil fuels coupled with environmental concerns has increased the interest in the research and growth of electric vehicle propulsion technologies. Automotive firms have been testing with different types of propulsion motors and energy conversion systems associated with advanced power conversion technologies. With this change in method the technical world foresees major advancements in technology giving birth to a product which will lead the global market in the coming years.

*Index Terms- Electric vehicle (EV), plug-in hybrid electric vehicles (PHEV), battery, electric motor, power electronics, propulsion.*

## I. INTRODUCTION

Electric vehicles (EVs) have been introduced in the late 1800s. They were very popular and a number of EVs were sold until about 1918. With the advancement of gasoline engines and their low-cost, and the discovery of electric starter for the ICEs, the interest in EVs completely dropped. In meanness of this, some companies continued to work on research and advancement of EV technologies by experimenting with different types of propulsion motors, energy storage systems, and power conversion technologies.

However, when the gasoline prices fell during the late 1970s, the EV activity again dropped. In 1980s because of the environmental concerns, the interest in electrical vehicle resumed. General Motors (GM) IMPACT (original prototype version of EV1 electric vehicle) was developed in early 1990s out of concern for air quality. The IMPACT design was based on advanced propulsion system technology and was designed suitable for mass production [1]. The EV1 was a battery-operated EV based upon lead acid batteries with induction motor as propulsion motor. Because of the restricted range of the lead acid batteries, in 1999 GM moved to nickel metal hydride (NiMH) batteries that offered a longer range but at a much higher cost. During 1990s, other automobile companies were also developing EVs.

In the last ten years, demand of plug-in hybrid electric vehicles (PHEVs) is increasing in North America and in other countries. This requires a relatively higher capacity battery compared with the typical state-of-the-art HEV batteries. GM's Chevrolet Volt, Toyota Prius, and Ford C-Max Energy are some of the PHEVs that are readily available in the marketplace. Other auto makers are also planning to commercialize PHEVs in the very near future.

The critical subsystem that is required in an EV is the propulsion system, which offers the tractive effort to drive a vehicle. The propulsion system in an EV consists of an energy storage system, the power converter, and the propulsion motor and related controllers. The battery is widely used as energy storage system and its charging is an integral part of the EV system.

## II. TYPES OF ELECTRIC VEHICLE

Electric vehicles broadly classified into three categories:

- **Battery Electric Vehicles (BEV):-** which is powered by a plug in charged battery. Electric motors have high torque at high speed and therefore do not generally require a gearbox. Generally, Electric motors have very high efficiencies about ~90+% compared to internal combustion engines about ~30-45%.

- **Hybrid Electric Vehicles (HEV):-** when petrol or diesel engine is used to generate electricity, which is used to powers the electric drive motor. Excess electricity is stored in a battery. Two types of hybrid drive vehicles are available, series and parallel hybrids:

  o A series hybrid is an all-electric drive-train which decouples the combustion engine from the drive shaft and allows the elimination of the gearbox.

  o A parallel hybrid operates the combustion engine in parallel to the electric motors and uses an automatic gear-box, the vehicle can run on electric, fossil fuel or combination of both.

- **Plug-in Hybrid Electric Vehicle (PHEV)**:- similar to the hybrid, excluding it uses a larger battery store to allow a portion of its energy to come directly from the electricity grid, recurring to petrol or diesel energy when the battery charge is down to a certain level [2].

## III. BENEFITS AND DRAWBACKS OF ELECTRIC VEHICLES

Electrical automobiles are propelled by an electrical motor powered by rechargeable battery combinations. Electric powered motors have own several advantages over internal combustion engines:

- **Energy efficient**- Electric engines convert 75% of the chemical energy by the batteries while internal combustion engines just convert 20% of the energy stored in fuel.

- **Eco friendly**- Electrical cars emit no pollution; but the power plant generating the electrical energy could produce them. Electrical energy by nuclear, hydro, solar, or wind-powered plants causes no air pollutants.

- **Efficiency benefits**- Electrical motors provide noiseless, smooth operation and high acceleration and require less maintenance compared to internal combustion systems.

Electrical vehicles face significant battery-related challenges:

- **Driving range**. Most electrical vehicles may just go about 150 miles after charging while gasoline automobiles can go over 300 miles after fueling.

- **Recharge period**. Fully recharging the battery combination can take up to 4 to 8 hours. Even a "quick charge" to 80% capacity may take 30 min.

- **Battery cost**: The battery combinations usually are expensive and might need to be changed one or more times in a year.

- **Bulk and weight**: Battery combinations are large and take up significant vehicle space.

## IV. ELECTRIC MOTOR

Electric motor converts the energy supplied by the battery into mechanical energy to provide power to the wheels. It is the combination of motor with the controller that controls the characteristics of the propulsion system and the choice of the power devices in the power converter. The main requirements for propulsion motors are roughness, greater torque to inertia ratio, high torque density, wide speed range, low noise, low or no maintenance, ease to control, and low cost.

Various types of electric machine technologies are being used for automotive propulsion. These include induction, switched reluctance, permanent magnet (PM) and axial gap machines. Most of the commercially available electric and hybrid vehicles use either induction or PM motors for propulsion. Automotive manufacturers and suppliers have significantly enhanced the electric machine technologies to be used in electric vehicles. Tesla Roadster induction motor peak power density is more than two times that of the induction motor used in GM EV1. Presently, the interior permanent magnet (IPM) synchronous motor is widely used in automotive propulsion because of its high power density, high efficiency and high torque. Toyota Prius, Ford Escape and Chevy Volt are some of the vehicles that use IPM machine [3].

The IPM-based motors are presently being used in most of the electric vehicles, there is a great concern about the availability of rare earth-based magnets and their increasing cost. A number of companies and researchers are working on the improvement of motors that don't use PM motors but to achieve the same performance as IPM motors. These include Switched Reluctance, Induction, Synchronous Reluctance, and PM associate Synchronous Reluctance Motors. In the coming future, the interior PM motor will likely continue to rule the market.

## V. POWER ELECTRONICS SYSTEM

Power electronics is enabling technology for the development of electric vehicle propulsion systems. The power electronics system consists of power switching devices, power converter topology with its switching strategy, and the closed-loop control system of the motor. The selection of power semiconductor devices, converters/ inverters, control and switching strategies and the system integration is very important for the development of efficient and high-performance vehicles. The challenges lie in obtaining a high-efficient, rugged, small size, and low-cost inverter and the associated electronics for controlling a three-phase electric machine. The devices and the other components need to withstand thermal cycling and extreme vibrations. All the current EVs and HEVs use a three-phase bridge inverter topology for converting the dc voltage of the battery to variable voltage and variable frequency to power a three-phase ac motor. Three-phase hard switched bridge inverter is the topology that is being used in all the electric vehicles. This topology is simple and well proven and continues to be the technology of future with various types of power devices and the corresponding passive components for filtering, electromagnetic interference (EMI) reduction, protection, and so on.

With the advancement of semiconductor device technology, various types of power devices with changeable degrees of performance are available in the market. Currently, IGBT devices are being used in most of the commercially available EVs, HEVs, and PHEVs. The IGBTs devices will continue to be the technology in near future until the silicon carbide (SiC) and gallium nitride (GaN) based devices are commercially available at a cost equally to that of silicon IGBTs. A significant progress is already made in the technology of these devices for automotive and other power applications [4].

The potential areas for the deployment of wide band gap devices in hybrid and EVs are: propulsion inverter, on-board battery charger (EVs and PHEVs), and the dc-dc converter for converting the high voltage to 12 V dc. These devices have lower conduction and switching losses, thereby offering higher efficiency in electronic systems. Use of these devices in the propulsion inverter reduces the weight and size of the unit because of the requirement for lower cooling for the same rating of the silicon-based power converter. In HEVs, it would be possible to combine the cooling of the power converter device and the motor with the engine coolant loop operating at 105 °C, thus reducing to one coolant loop leading to further decrease in weight and complexity.

In addition to the power conversion and control of propulsion motor, observing the conditions of the electric machine is very important in EVs to sense any failures such as bearing, stator and rotor faults. By detecting the electric machine faults as soon as possible, the lifetime of an electric machine can be prolonged by performing maintenance before a disastrous failure occurs. Therefore, the EVs require embedded fault diagnosis systems both to provision critical functions of the control system and to deliver an economical maintenance [5]. Unless the electric machine and power train components are continuously observed, motor faults might cause permanent damage depending on the severity of the fault. The prognostics and health management are presently not being implemented in most of the EVs and HEVs. Integration of prognostics in the overall control system could expect the future performance of the machine by assessing extent of its deviation from its expected normal operating conditions.

## VI. ENERGY STORAGE SYSTEM

EV use is limited because of a restricted range, increased refueling (or recharging) time and cost. These aspects are closely related to the energy storage system. The main parameters in the selection of a battery for EV applications are: power density, energy density, weight, and volume, cost, and life cycle. The other parameters are operating temperature range, material recycling, safety and maintenance. Power density determines the acceleration ability. Energy provides an indication of the potential range. Life cycle measures how often the battery can be recharged to its full capacity and related to battery lifetime. Weight and volume can affect the range as well as efficiency of the total system. Cost is determined by the availability of resources, technology, and manufacturability. Since EV is driven by environmental concerns, the recycling capacity of the material is also an important consideration. In EV applications, to operate at higher voltages, several battery units have to be connected in series. Thus, safety, charge equalizing, and reliability become challenging problems to be addressed.

Until the mid-90s, almost all the EVs used lead acid batteries although some vehicles used other kinds of batteries such as nickel cadmium batteries. The GM EV1 was using valve-regulated lead acid batteries at 312 V and in 1999, switched to NiMH batteries, and Toyota Prius also uses these batteries. These NiMH batteries are used till recently in almost all the commercial electric and hybrid vehicles. Tesla Roadster is the first production automobile to use lithium-ion battery and the first production EV to travel about 200 miles/charge. Presently Nissan leaf EV and GM's Chevy Volt plug-in EV also use lithium ion batteries.

Lithium-based technologies and lithium ion batteries are leading the way to meet the requirements of EV/HEVs. These batteries can give high energy and power per unit of battery mass, allowing them to be smaller and lighter than other rechargeable batteries. Other advantages of lithium-ion batteries compared with lead acid and NiMH batteries include high energy efficiency and a relatively long life cycle. Lithium-ion is clearly a better and more effective way to power modern hybrids and EVs, but it is presently more expensive.

The future of EV battery could be based on lithium air technology. These batteries could significantly increase the range of EVs because of their high energy density, which could be equal to the energy density of gasoline. Scientists estimate that these batteries could hold 5 to 10 times the energy of lithium-ion batteries of the equal weight and twice the energy for the same volume [6]. They have the potential of achieving the energy density in the range of 2000-3500 Wh/kg. No other known battery has as high of an energy density as lithium-air batteries. These batteries have an anode of lithium and an air cathode. When the lithium combines with the oxygen, it forms lithium oxide and releases energy. As the oxygen does not need to be stored in the battery, the cathode is lighter than that of a lithium-ion battery, which gives lithium-air batteries their high energy density.

## VII. BATTERY CHARGING

It is well known that we will never be able to charge the battery of an EV as fast as filling the tank of an automobile with gasoline. Current plug-in and EVs are designed primarily for home charging which charge using ac supply. These charger units that use 120- or 240-V ac are generally installed on the vehicle (on board charger). Home owners must install electric vehicle supply equipment (EVSE) to connect home energy management system (HEMS) with the on-board chargers. Other chargers are off-board and use dc charging, and the term dc fast charging is mostly used to refer to these chargers [7].

Several companies are already developing EV smart charging, which is the integration of energy flow and information flow. To date, the design and development of EVSE are focused on power protection, security, and billing functions. When PHEVs and EVs are connected to the grid, its integration into the distribution infrastructure needs to be managed through bidirectional communications. The charging infrastructure would also improve by having local wireless network architecture with connectivity between the home area network gateway and EVSE that will minimize the communication requirements and cost of the EVSE. The EVSE to EVSE communication networks and neighborhood area networks that do not require any direct connection to the utility company would be the way of future for the EV charging system [8].

The convenience of charging could be a major factor in purchase decisions of the EVs. All of the major EV manufacturers have announced partnerships for developing the technology to address the concern of range anxiety. A number of companies are developing inductive charging that uses an electromagnetic field to transfer energy for charging the batteries. This type of wireless charging eliminates the EV power cord with an automatic charging solution. Automakers such as BMW and Nissan are already implementing wireless charging options on their electric cars, which could allow for charging locations to be embedded in parking spaces and even the roadway. For example, a wireless charging system is being developed by Delphi that will automatically transfer power to a vehicle providing a convenient, wireless energy transfer. This hands-free charging technology is focused on highly resonant magnetic coupling that transfers electric power over short distances without physical contact, allowing for harmless and more appropriate charging options for consumer and commercial EVs [9].

## VIII. CONCLUSION

The advancement of propulsion system technology of EVs will be focused on five areas: 1) vehicle range; 2) vehicle cost; 3) battery pack replacement cost; 4) battery pack life; and 5) quick and easy recharging. Apart from vehicle cost, most propulsion system development work will be on battery systems. Significant improvements in lithium ion technology have been achieved and these batteries are being installed in several EVs and PHEVs. More of the PHEVs and EVs based on lithium ion battery will be available from various automakers in the nearby future. A large research effort is going on to develop lithium air batteries for automobiles. The advancement of lithium air battery will improve the range of PHEVs and may lead to dominance of pure PHEVs.

Dc fast charging of PHEVs and EVs is projected to increase in the future. The smart charging features with the integration of energy flow and communications will be combined in all the chargers.

Next-generation EVSE must adjust to become smarter and more capable in future. It must support smart-grid functionality, which holds great potential to lower the system cost of providing energy to PHEVs and EVs. With smart-grid interactions, not only the distribution grid will be better optimized for lower energy costs but also the users can use the PHEVs and EVs at lower electricity costs, and charge faster with higher efficiency.

### REFERENCES

[1] K. Rajashekara, "Present Status and Future Trends in Electric Vehicle Propulsion Technologies", IEEE Journal of Emerging and Selected Topics in Power Electronics, 2013

[2] A guide to electric vehicle, available at http://www.seai.ie/Your_Business/Technology/Industry/Electrc_Vehicles.pdf

[3] (2013, Feb.). Electric Motors for Electric Vehicles 2012-2022 [Online]. Available: http://www.idtechex.com/research/reports/electric-motors-for-electric-vehicles-2013-2023-forecasts-technologies-players-000344.asp

[4] M. Kanechika, T. Uesugi, and T. Kachi, "Advanced SiC and GaN power electronics for automotive

4

systems," in Proc. IEEE Int. Electron Devices Meeting, Dec. 2010, pp. 1–4.

[5] B. Akin, S. B. Ozturk, and H. Toliyat, "On-board fault diagnosis of hybrid electric vehicle motors at start-up and idle mode," IEEE Trans. Veh. Technol., vol. 58, no. 5, pp. 2150–2159, Jun. 2009.

[6] Why the Lithium Air Battery is Over Hyped [Online]. Available: http://gigaom.com/cleantech/why-the-lithium-air-battery-is-over-hyped/

[7] K. Morrow, D. Karner, and J. Francfort, "Plug-in hybrid electric vehicle charging infrastructure review," Dept. Energy Vehicle Technol. Program, Univ. Advanced Vehicle Testing Activity, New York, Ny, USA, Tech. Rep. INL/EXT-08-15058, Nov. 2008.

[8] M. Erol-Kantarci and H.T. Mouftah, "Prediction-based charging of PHEVs from the smart grid with dynamic pricing," in Proc. IEEE 35th Conf. Local Comput. Netw., Oct. 2010, pp. 1032–1039.

[9] (2010, Sep. 29). Delphi Working to Make Electric Vehicle Wireless Charging a Reality [Online]. Available: http://www.witricity.com/pdfs/Delphi_Press_Release_(9-29-2010).pdf

**Vaibhav Hudda** was born in Hapur, India, in 1991. He received the B. Tech (Electrical and Electronics) degree from Subharti Institute of Technology and Engineering, Meerut. (affiliated to UPTU, Uttar Pradesh), in 2011 and, currently pursuing his M. Tech (Electrical Power and Energy Systems) from Ajay Kumar Garg Engineering College, Ghaziabad (affiliated to UPTU, Uttar Pradesh).

**Akshay Kumar** was born in Ghaziabad, India, in 1988. He received the B. Tech (Electrical and Electronics) degree from Ideal Institute of Technology, Ghaziabad (affiliated to UPTU, Uttar Pradesh), in 2009 and, currently pursuing his M. Tech (Electrical Power and Energy Systems) from Ajay Kumar Garg Engineering College, Ghaziabad (affiliated to UPTU, Uttar Pradesh). He worked in Industry and he has also teaching experience of more than 1 and half year in Engineering College.